Be Careful in Florida

Know These Poisonous
Snakes • Insects • Plants

Written and Illustrated by

Francis Wyly Hall

Author of

- Shells of the Florida Coasts
- Palms and Flowers of Florida
- Birds of Florida

GREAT OUTDOORS PUBLISHING CO.

4747 TWENTY-EIGHTH STREET NORTH   ST. PETERSBURG, FLORIDA 33714

Florida

〰

Florida is a wonderful land. A land of milk and honey, palms, pines and giant oaks, restless surf and quiet lakes. Here in the warm, sunny climate, flowers and plants spread a riot of color against thick, green leaves, and all sorts of animal life thrive contentedly. To quote Marjorie Kinnan Rawlings, Florida is a "Great and beautiful tropic queen."

Looking beneath this placid surface we find that nature, while always wonderful is not always beneficent to mankind. We find that nature has provided most of her children with a means of protection. Snakes can bite, and insects can sting. Many lovely, flowering plants contain dangerous poisons, and other are armed with spines. It is with these dangers that this little book deals. Not that it has been written for the purpose of alarming anyone, nor is Florida any more dangerous than other places, but we do have many visitors to this state, who are not familiar with the local flora and fauna. Nearly everything described on the following pages can be found in any of the southeastern states. We all know that poisonous snakes are dangerous, but with a minimum of precaution you will never be bitten by one, and if you know something about snakes you will not be afraid of them. Plant poisoning frequently causes the death of livestock. Only very rarely does it cause the death of a human being, except in the case of poisoning by mushrooms. It does happen occasionally though, and oftener people are made ill. Wise parents will teach their children the serious danger of chewing berries, seeds, or roots from unfamiliar plants, as these parts often contain more concentrated poison than the rest of the plant. It doesn't matter if you are in Florida, Virginia or Illinois, as poisonous plants are found in all sections of the country.

However, to be forewarned is to be forearmed, and so this book is written to pass along to you the knowledge that will prevent accidents, and unpleasant experiences.

Poison is often not a death-dealing agent. It is something that causes harm, or pain, or illness, and if taken in large enough doses, can be fatal.

∽

Webster's Dictionary gives the following definitions:

Poison: Any substance noxious or harmful to life or health; that which taints or destroys purity or health; venom.

Poisonous: Having the qualities of poison; impairing soundness or purity; corrupting.

Snakes Can Kill

Someday your life may depend on whether you can tell a poisonous snake from a harmless one. Of all the 2,000 species of snakes native to the United States, only four species are poisonous, and they can all be found in Florida. Fortunately snakes are shy, and will gladly retreat and let you go on your way unharmed if given the opportunity. Nearly all will fight if frightened or cornered, and, of course, they don't like to be stepped on. In reality snakes are among man's best friends! They keep our rat population under control, and these rodents do a vast amount of damage to crops and buildings, and carry some of the worst diseases known to mankind. Also many harmless snakes will kill or drive away the rattlers and moccasins. Therefore, it would be wise to know the poisonous snakes, and kill those, but let the harmless ones go free. As a matter of fact the harmless snakes make fine pets. They are exceptionally clean and become tame and gentle.

Florida's poisonous snakes are Rattlesnakes, Copperheads, Cottonmouths, and Coral Snakes. The first three are known as pit vipers, because they have a deep pit or depression between the eye and nostril. They have vertical, slit-shaped pupils in the eyes, where the harmless snakes have round pupils. These three have long fangs, or hollow teeth in the front of their mouths, which inject the poison much in the manner of a hypodermic needle. These fangs fold back into the roof of the mouth when not in use. Every three or four months the old fangs are shed and new ones take their place. Also, snakes shed their skins several times a year, exchanging the old ones for shiny, new outfits. They have flat, triangular heads, distinct from the neck, and their scales have a keel-like line running from base to tip. They do not lay eggs, but give birth to live young.

The Coral snake belongs to the Elapine family, of which the deadly Cobra is a member, and they are quite different from the pit vipers. *See Coral snake.

Do not let the thought of poisonous snakes keep you from enjoying any of Florida's beauty spots. I have seen very few of these serpents in their natural haunts, and those I have seen, were intent on minding their own business, and we didn't disturb each other.

5

CORAL SNAKE

COTTONMOUTH
MOCCASIN

DIAMOND-BACK
RATTLE-
SNAKE

PIGMY
RATTLE-
SNAKE

COPPERHEAD

Florida Rattlesnakes to watch out for are the Diamondback, Canebrake (Seminole or Swamp), and the Pigmy (Ground) Rattler. They all have the distinguishing rattle at the end of the tail.

The rattle is made up of cup-shaped joints that fit loosely into each other, with a new joint being added each time the snake sheds its skin. With age the joints wear and the end ones break off. The snake usually gives a warning rattle, and coils before striking, hissing and trying to warn the intruder away. If it does strike, the movement is so lightning quick, it can barely be followed by the human eye, and it can strike with great accuracy for one-half of its length. The reptiles can be found almost anywhere and do not mind a swim in either fresh or salt water. In winter they will lie in the warm sunshine, but during much of the year the sun is too hot for them.

Diamondback Rattlesnake (Crotalus adamanteus). This is our largest snake, growing up to 8 feet long, and as big around as a man's upper arm. Heavy body is plainly marked with black diamonds bordered with white. Body is grayish or brownish yellow, lighter beneath. Often the coloring is olive and the diamond markings are bordered with yellow. There are paler less regular markings on the sides. The Diamondback is beautiful in its markings and coloring especially when the skin is new. This is also the time when his temper is apt to be the shortest.

Pigmy Rattler (Sistrurus miliarius) or Florida Ground Rattler. A small member of the clan, though stout of body, growing to 2 ft. long. The flat head has shields over the eyes. Color is gray-brown, with a series of black blotches on the back, and lighter less distinct ones on the sides. The belly is a dirty white with dark spots, and there is a reddish tinge along the back and on the tail. The small rattle can only be heard a few feet away. These Pigmy Rattlers are more common than people think. Their bite is proportionately less severe than that of the larger members of their family, because of their smaller fangs.

Canebrake Rattler (Crotalus horridus). Also called Swamp, Seminole, or Banded. A heavy-bodied reptile, 5 ft. to 7 ft. long. In coloration and pattern this snake shows much variation. Florida specimens are usually grayish or olive with a pink tinge, or brownish pink. A rusty-reddish streak runs along the back and on top of the head, with a dark band on the side of the head. It is marked with dark cross-bands, that are irregular, belly pinkish or yellowish, and a black tail.

Copperhead (Ancistrodon contortrix) or Highland Moccasin. The adult Copperhead is 3 ft. to 4 ft. long, with distinctive, hour-glass shaped bands. General color is brown or gray, and the bands are dark chestnut. The head has a coppery tinge. Belly is pinkish-white with large, dark spots along the sides. Young specimens have the tail tipped with bright, yellow-green. There are relatively few Copperheads in Florida, and the serpent is not antagonistic by nature, so they rarely cause trouble.

Cottonmouth Moccasin (Ancistrodon piscivorus) or Water Moccasin. This reptile has a very large, heavy body tapering suddenly to a thin, pointed tail. It attains a length of 5 ft. General coloring olive-greenish or brownish, with irregular, slightly darker bands. Under side is yellowish, blotched with dark. The top of the head is nearly black, upper lip plates yellow and yellow bars on the lower lip and chin. Shield-like scales project over the eyes. This is an ugly fellow, both in appearance and disposition. Give him a wide berth as he is very venomous. Often seen around streams, ditches, swamps and lakes, maybe sunning himself along a limb or on a bank. When annoyed or too closely approached the moccasin will throw back its ugly head, and open its cottony mouth, wide, with the fangs erect, ready to strike. They can bite just as well under water. It resembles several harmless water snakes, but they are very timid and will hurry off when anyone comes near. The Cottonmouth seems to be more aggressive and antagonistic than the other poisonous snakes.

Coral Snake (Elaps fulvius) or Harlequin Snake. This beautiful, deadly serpent, grows to be 2 to 2½ ft. long. It is slender, smooth-scaled, with a small, blunt head and beady eyes with round pupils. The fangs are short and permanently erect. It is gaudily colored, with alternating black and red bands around the body, with a yellow ring between each band. The snout is black, the rest of the head yellow, then black, yellow, red, yellow, black, yellow, etc. There are several harmless snakes that may be mistaken for the Coral at a glance but none of these have the yellow rings separating the red and black bands. A well known jingle is:

"Red touch yellow, kill a fellow—
Red touch black, good for Jack"

This reptile is found throughout Florida. Very few people are bitten as it is shy and usually stays hidden during the day, coming out after dark to prowl around and feed. It is good-natured as a rule, and not apt to bite unless frightened, provoked, or restrained. The Coral Snake does not coil before it strikes. If annoyed it will twist from side to side with jerky movements. Then it is dangerous and as quick as lightning. It chews, rather than bites, embedding its short fangs several times. The venom is extremely poisonous, and works very rapidly. Heavy clothing is good protection against the short fangs of this reptile. Coral Snakes lay several long-oval eggs, which hatch in three months. The young are 7 inches long, very active, and will try to bite almost from birth, if alarmed.

Prevention and Treatment of Snake Bites

Use a little common sense if you are tramping about the fields and woods. Stick to paths where possible, or at least avoid clumps of underbrush, as snakes like to hide under bushes or behind logs. Wear heavy boots or shoes and long, heavy socks, with pants that come down well over the ankles. Carry a stout stick, and don't make like an Indian, but make lots of noise and beat at the bushes as you go along, and all snakes in your vicinity will keep well out of your way. If you must poke into hidden places, do so with the stick and not your hands, as most snake bites are received on the hands or ankles. Keep a sharp lookout for the Cottonmouth near streams, ditches, ponds and such places. Also be cautious about going into long unused buildings or barns. Snakes think that old piles of trash or wood, stumps and hollow trees are put there just for them. Be Careful!

If someone does get bitten by a snake, and you are not sure it was a poisonous snake, examine the bite. Keep these two drawings well in mind and you will be able to tell the difference. Non-poisonous serpents can inflict a small bite, which should be dabbed with antiseptic. For a bite by a venomous snake, put a tight, constricting band around the limb of the victim, above the knee or elbow. This should be done within 30 minutes. With a sharp knife or razor blade, make a cut over the fang mark and along the limb,* ¼" deep, and apply suction, to make the wound bleed. If the mouth is used for this purpose, be sure there are no cuts or open sores in it, and be careful to spit out all the blood. When possible have the victim suck his own wound, as he is already contaminated with the poison. As the limb swells, loosen the tight band, as it is dangerous to stop all circulation. Get the victim to a doctor as soon as possible, but he should NEVER run, and should not walk unless it is necessary. The venom of the pit viper attacks the blood stream, so anything that increases circulation will lessen the chance of recovery. Do not give stimulants, or alcohol in any form, but applying ice or immersing the limb in ice-water is helpful. If a snake-bite kit, containing anti-venom is available, inject 50 centimeters in the tissue around the bite. This will greatly improve the chance for survival.

It is advisable to take a snake kit with you if going into wooded sections, but the anti-venom must be kept refrigerated unless used within 24 hours. Another method now being used is to "freeze" the area around the bite with Ethyl Chloride. A tube or bottle may be obtained from the drug store, with instructions for use, and

---

* Do not use the cross-cuts, as they are apt to cut a nerve.

taken along on a trip. Ethyl Chloride is extremely volatile and care must be taken to keep it in the shade, and not to drop it, or it could explode. Also it is easy to use too much and "burn" the victim. The freezing process does slow down the circulation and allow more time to get to a doctor. It is still wise to apply the constricting band at once, make the cuts, and suck out as much blood and poison as possible, before applying the Ethyl Chloride.

There is now an anti-venom for Coral Snake bite, but it is not available everywhere, and must be given promptly to be effective. Call the police or local Health Department for information.

Symptoms — Pit vipers: severe pain, swelling and dark discoloration of the skin, weakness, shortness of breath, dimness of vision, nauseau and vomiting, rapid pulse and unconsciousness. The venom of the Coral snake acts on the nervous system. There is a burning pain in the wound, followed quickly by weakness, loss of speech, spasms, labored breathing, coma and death.

Do not become panicky. If first aid treatment is given at once, and the patient then taken to a doctor or hospital, the bite will probably not prove fatal.

～♍

Bites of poisonous and harmless snakes, showing the teeth marks.

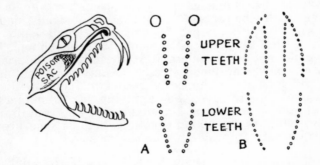

A—Shows the teeth of the Pit Viper. The two large holes would represent the fang punctures.

B—Shows the bite of the non-poisonous snake. There are N O f a n g holes.

Insect Stings and Bites

The commonest type of sting is that of bees, wasps and hornets. These usually cause only painful swelling, with redness, heat and itching, for a short time. If several of these stings are received at the same time, the victim may become ill. Some people are extremely allergic to bee stings, so much so that even a single sting can cause death. If the victim develops labored breathing or shortness of breath, he should be rushed to the nearest doctor or hospital. This is very rare, so don't be alarmed over a bee sting.

Ants, of course, can be found everywhere. There are some very small, red ones, called Fire Ants, which can, as the children say, "Sting the fire out of you." There is a campaign under way to eradicate them in Florida. The Velvet Ant is a very large one — with a painful sting.

Ticks are numerous in some areas. It is wise to inspect children who have been playing outdoors, every night. Most ticks only suck a little blood, but there is always the possibility that one might carry the dread Rocky Mountain Fever, although it is rare in Florida. Put a little turpentine on the tick, and it will loosen its hold enough to be pulled off with a pair of tweezers. If just pulled loose the tick if deeply embedded, will leave its head in its victim's flesh, and this will cause a bad sore.

Black Widow Spider. The body is very round, less than ½ in. long, and shiny black, with a bright red patch on the under side. She spins her web in some secluded, dark place, mates, devours her husband, and lays several hundred eggs in a cocoon. The eggs are extremely toxic and should not be handled. This is one of the most poisonous creatures known, the venom being many times more powerful than that of the Rattlesnake. Symptoms are severe abdominal pains, due to muscle spasm, dilated pupils, swelling of face and extremities, collapse and convulsions. First aid treatment is the same as for snake bite, and there is an anti-venom, so in all cases get the patient to a doctor or hospital at once. Most victims who receive quick treatment recover. The spider is not vicious or aggressive, and will not bite unless provoked or frightened by being captured.

The only other spider whose bite is poisonous, is the Tarantula. This big, hairy fellow has a bad reputation, which is much worse than he deserves. The bite is only mildly poisonous to humans, and they are voracious feeders on insects. All spiders are very beneficial to mankind, devouring as they do, vast quantities of roaches, bugs and plant pests. In my house I never kill a harmless spider, or snake.

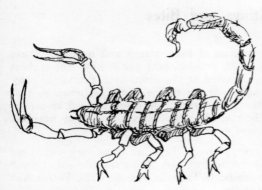

Scorpion. The Scorpion which is a member of the spider family, looks somewhat like a miniature lobster, with a long, slender tail which is divided into sections. They are dark brownish, and are 2 in. to 3 in. long, not including the tail. The Scorpion administers its sting from the tip of the tail, which is raised forward over its head, and struck down. There is a bulb containing the poison, and from it extends a sharp, curved barb. The creature will run away if possible. People are usually stung when reaching the hand into a dark corner, behind books or boxes, into a kitchen cabinet, or behind something in a garage. The sting of the Florida scorpion is not fatal, but is very painful, and in severe cases can cause temporary shock and coma. There is a burning sensation at the site of the sting, pain over the entire limb, followed by headache, dizziness and nausea. It is best to call a doctor.

There are several other insects to be careful of. These include Lo Moth Caterpillar, a fat, green fellow, with a red stripe along his back. The black-tipped hairs growing in little bunches on his body, are mildly poisonous.

Saddleback Catterpillar has a reddish or orange saddle, on a green saddle pad on his back. The body is brown, and it has the appearance of having a head at each end. It has spines which sting like nettles. Usually found on fruit trees.

The Assassin Bug is a curious looking beetle, 1½ in. long, with a notched crest on his head like a cock's comb. He can bite, injecting a saliva-like poison, causing a painful wound.

Centipedes are found throughout the world, but only those of the tropical regions are venomous. Their bite is fatal to small animals and is painful to man. The body of the Centipede is divided into segments, each bearing a pair of legs. The number of segments varies in different species, from 15 to 150 or more. They usually live in damp places under stones or timbers. They are night feeders, and are timid.

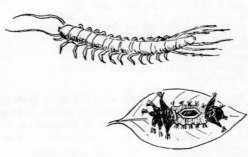

There is also a curious little chap called a Puss Moth Caterpillar. He is soft and fuzzy looking, cocoa-brown in color, with delicate, feathery things, maybe feelers or antennas, waving from the top of his head and along his back. Be careful, and don't touch! He'll cause a severe, burning rash to break out on the skin.

Stings of Marine Animals

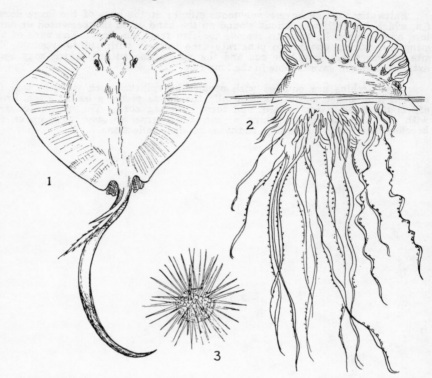

Portugese Man-of-War. If you see a lovely, iridescent blue balloon floating by, on or just under the surface of the water, be careful. From the under side of this balloon stream a mass of fine tentacles, sometimes 30 ft. long, along which are stinging cells, each of which can sting like a wasp. When washed ashore, and for some time after death, these tentacles can still sting. They usually seem to come by "spells" and when, for a day or so, the jellyfish are seen along the beach, bathers should be careful. Along the Atlantic Coast, a strong, easterly wind will blow them inshore, from the Gulf Stream. There are other jellyfish than can sting, too, so just give the whole family plenty of room. Soak with dilute ammonia water, followed by hot epsom salts. Occasionally, if the shock is extreme, a doctor should be consulted, but usually home remedies will suffice.

Sting Ray or Whip Ray. This sand-colored creature lies on the bottom of seas, bays or rivers, its flat body partly concealed by sand or mud, making it difficult to be seen. When stepped on it flips the long tail up and around and drives the saw-toothed spine into the leg or foot. The pain is intense, aggravated by some venomous substance on the spines. Swelling and discoloration of the limb follow, with dizziness and nausea. Symptoms of prostration and blood poisoning may develop, requiring hospitalization. If a doctor cannot be reached promptly, use the same first aid treatment as for snake bite. These fish never attack anyone, and will take to deep water if several bathers are splashing around. The common Skate, which is similar in appearance, and the King or Horseshoe Crab are entirely harmless.

Saltwater Catfish have a venomous stinger at the base of the large dorsal fin, which can inflict a serious wound on the hand of an inexperienced or careless angler when he tries to remove the fish from the hook. The barb of the stinger is strong enough to penetrate shoe leather, with painful results. The site of the sting should be cut and the poison sucked out, which may seem extreme, but will save trouble in the long run.

Sea Urchins are equiped with needle-like spines, which wave to and fro on ball and socket joints. Occasionally the points pierce a bather's leg, break off in the flesh, work inward, and form festering sores. The urchins, covered with dark reddish-purple spines, or sometimes greenish ones wash up on the beaches and make interesting additions to shell collections.

Poisonous Plants

◦◦◦

There are two types of plant poisoning. The toxic material may poison the skin, causing a rash, swelling, and severe burning or itching. In the other type the poison enters the digestive tract. The first is limited almost entirely to humans, while the second is more likely to cause trouble for animals, as they are more apt to eat the plants.

Many plants are not dangerous under all conditions. Some are toxic only in the seedling stage, others when the plants have matured, or it may be while the berries are green, or ripe. Often only certain parts of a plant are harmful, or a small amount will do no damage. The flowers or foilage of some of our loveliest garden favorites contain poison, but are not dangerous because they are seldom or never eaten, such as Larkspur, Clematis, Lupine, Hydrangea, Laurel, Foxglove, Jack-in-the-Pulpit, and many lilies which have poisonous bulbs, especially Amaryllis. Iris, Daffodil and Narcissus all have poisonous root stocks. The plants with berries or seeds which are toxic, are more apt to cause trouble, since children are tempted to eat them.

Some plants cause skin poison to only a few very sensitive people, such as Buttercup, Primrose, Rue, Arnica and Queen Anne's Lace. These may also cause inflamation of the eyes, if you should rub them after picking the flowers. Each section of the country has various kinds of stinging nettles, usually with attractive flowers, which can be recognized by the little stinging hairs on the stems and leaves.

There are also plants that wound, such as Castus, Yucca, Sandspur and Nettles.

A great many of the wild cherries, such as Laurel Cherry, Wild Black Cherry, Carolina Cherry, etc., and also the Black Locust and the Tung Tree have a dangerous, poisonous principle in the leaves and bark. Trimmed branches from these trees should not be left where stock might eat them. Berries of the Holly tree are toxic, and should never be eaten.

Poisoning of stock from plants is a much more serious problem than most people realize. The animals often die from the amount of poison consumed.

This is a small book, so I have selected only the most poisonous plants, or those that might cause trouble to humans or domestic animals.

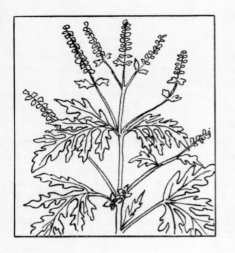

Ragweed (Ambrosia artemisiifolia). An annual, persistent weed; erect, much branched, rough hairy stem; 1-5 ft. high; leaves mostly opposite, smooth, thin, lightish-green, deeply cut into segments and pointed lobes; slender spikes of insignificant, greenish flowers grow from the axils of the leaves; these are followed by tiny seed pods containing a single seed. This weed is found in fields, yards, gardens, along roadsides, and just about everywhere. While it is not really a poisonous weed, it causes misery to many people, for Ragweed produces prolific quantities of wind-borne pollen, which is the cause of most cases of late summer and autumn hay-fever. Plants that cause hay-fever have pollen that is light, plentiful, dry and toxic, and carried by the wind, not by insects. Such pollen comes from grasses, oaks, alders, maples, elms, and pigweeds and ragweeds.

Plants That Cause Dermatitis

Dr. Walter C. Muenscher, in his book "Poisonous Plants of the United States," lists a hundred known to cause dermatitis in susceptible people. There are not many in Florida that seem to cause trouble. By far the worst trouble-makers of all, belong to the genus Rhus, which includes the Poison Sumac, and the poison oaks and ivies, which are neither oaks nor ivies. The characteristics of the plants in this group are variable, and botanists are not entirely agreed as to classification. They are all very bad company and should be severely left alone. They are all poisonous to most people, and often persons who have believed themselves to be immune, will develop a bad case of poison ivy. These plants secrete urushiol, a yellowish oil which rubs off on clothing and exposed skin. The toxic oil is carried to individuals in the smoke from burning underbrush. A preventative is to cover the exposed skin with a paste of baking soda and cold cream, if going into areas where you may encounter poison ivy. If exposed, wash the skin as soon as possible, in water in which a box of baking soda has been dissolved, then wash with yellow laundry soap several times in succession, and rinse well. All clothing should be thoroughly washed. Calamine lotion will help to relieve the intolerable itching, but in severe cases see a doctor.

BE CAREFUL! DON'T TOUCH!

Poison-wood or Coral Sumac (Metopium toxiferum). Large shrub or small tree; native to So. Florida and Keys; low-spreading crown of dark, glossy, broad leaves; orange, berry-like fruit; red-brown, flaky bark has milky juice.

Poison Sumac (Rhus vernix). North Florida in swampy ground; shrub or small tree; stout, spreading branches; gray bark; pinnate leaves with 7-13 pointed leaflets; small, whitish flowers in loose panicles, and small, yellowish-white, berry-like fruits. Leaves turn bright scarlet in the fall, and are tempting to pick.

Poison Ivy, Poison Oak, Poison Creeper (Rhus radicans). This is the common climbing form of the eastern states. Shrub or vine; stems woody; glossy or dull green leaves, in groups of threes, pointed at tip and base, and broadened in the middle, with irregular edges; the terminal leaf is on a short stalk while the two side leaves grow directly from the stem. Small, greenish-white flowers in open panicles; clusters of tiny, waxy-white, berry-like fruits. All parts of this plant are poisonous, even the dead leaves. The toxic oil causes inflammation and swelling, intense irritation and itching, followed by small blisters. Scratching aggravates the condition and causes it to spread to other parts of the body.

Manchineel (Hippomane mancinella). So. Florida and the Keys; short-trunked tree with drooping branches and a low crown of dark, rounded, evergreen leaves; fruits resemble small crab apples. This is one of the Spurge family, a widely poisonous group of plants. Even rain water dripping from the leaves can be injurious, and the sap affects both internal and external tissues. Most of these trees have been removed from populated areas, but they are still found in overgrown woodlands.

17

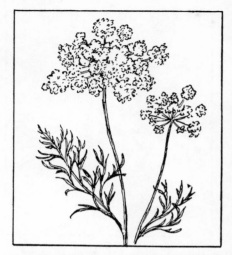

Snow-on-the-Mountain (Euphorbia marginata). Cultivated for the ornamental white-margined leaves, borne on branches at top of erect stem; 2 ft. tall; terminal clusters of tiny, white, insignificant flowers. This plant has a milky juice that causes dermatitis, as will touching the leaves, in some cases. The leaves are also poisonous if eaten. Another of the Spurge family.

Queen Anne's Lace, Wild Carrot, Bird's Nest (Daucus Carota). Herb, 2-3 ft. tall; erect, branched, hairy stems; compound leaves, deeply cut; tiny, white flowers in flat-topped lacy heads; strong odor attracts bees and insects. Dermatitis is caused in some cases, especially if leaves are wet, and be careful not to rub the eyes, as juice stings and burns.

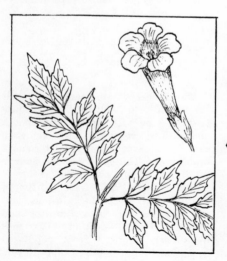

Tread-Softly, S p u r g e   N e t t l e (Jatropha stimulosa). Usually low to the ground; broad, lobed leaves on stalks; pretty, dainty, white flowers, ½ in. across. Common on dry, sandy soil. This plant is well covered with bristly, stinging hairs, which cause severe inflammation and itching. Belongs to the Spurge family.

Trumpet Creeper (Campsis radi-cans). A beautiful native vine, that climbs by aerial roots over fences and bushes and into low trees. Vine woody, widely branched; leaves opposite, compound, of 7-9 toothed, pointed leaflets; showy flowers in terminal clusters, red with yellow throats, trumpet-shaped, 2-3 in. long. Just be a little careful in gathering this beauty, as it causes dermatitis in some people.

BE CAREFUL! DON'T EAT!

Datura, Devil's Trumpet, Jimson Weed, Thorn Apple (Datura stramonium). Rank, herb-like; smooth, stout stems; leaves large, ovate-angular, coarse-toothed edges; beautiful, trumpet-shaped flowers, white or lavender, 4 in. long, often double; fruit a hard, spiny, oval pod, 2 in. long. This pod contains many seeds which are very poisonous. The alkaloid hyoscyamine occurs in the seeds, leaves and roots. The flowers and leaves are toxic even when dried. Children have been poisoned from eating the unripe seed pods.

Nightshade Family

The Nightshade Family includes several poisonous members as well as others that are harmless. The best known are the tomato, eggplant, and Irish potato. They are herbs, with angular, often hairy, much branching stems; alternate leaves that have a strong odor; regular, perfect flowers, with 5-lobed calyx and corolla; fruit a berry.

Irish Potato (Solanum tuberosum). The common potato, long-time staple of our diet, contains the poison solanine, in the eyes and new sprouts. Tubers growing at the surface of the soil, and exposed to sunlight, turn green. These "greened" tubers should always be thrown away, as humans have been fatally poisoned by them, and animals have died from eating new sprouts.

Jerasulem Cherry (Solanum Pseudo-Capsicum). A popular house plant, grown for its ornamental, bright orange, cherry-like berries. The toxic berries contain solanine and several other poisons, and should never be eaten. Warn the children about this one as the berries are very tempting looking.

Black Daytura (D a y t u r a meteloides). Very similar shrub, with white or yellow, or purple tinted flowers, and a round, spiny pod, 1¼ in. diameter. It has the same poisonous qualities as the Datura stromonium.

Ground Cherry or Love Apple (Solanum aculeatissimum) Woody plant of sandy fields and roadsides; stems have many sharp spines; red or orange papery berries, often used in dried arrangements.

Black Nightshade or Poison Berry (Solanum nigrum). Annual herb; erect, smooth, branched stem; 1-2½ ft. high; leaves ovate, wavy-toothed, broader at the base; little white flowers in small, side clusters; round, black berries are borne on drooping stems. Leaves and unripe berries are poisonous. Green berries resemble miniature tomatoes, and are tempting to children. Ripe berries are cooked and eaten in pies.

19

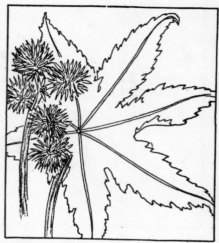

Oleander (Nerium oleander). Very handsome, widely grown shrub, to 20 ft. tall; very slender leaves, 7-9 in. long, smooth, leathery, evergreen; fragrant flowers, in close, terminal clusters, white, red, pink or peach. Every part of this plant is dangerously poisonous, green or dry, and can cause death to humans or livestock. In Italy it is known as Funeral Flower. Children should be forbidden to chew on the twigs or leaves.

The Oleander belongs to the Dogbane Family, which has several other poisonous members, among them the beautiful **Allamanda** and the **Peri-winkle**. Allamanda (Allamanda cathartica) prefers to grow as a vine, has gorgeous, large, yellow flowers and long-ovate, pointed, glossy, yellow-green leaves. The Periwinkle (Vinca rosea) abundant in Florida, is an erect herb, 24″ tall; glossy, oval leaves, and bears rosy-pink or white blooms during most of the year. The flowers of both plants have five broad flat petals, and the stems, when broken have the acrid, milky juice that is characteristic of the Dogbane. The Allamanda and Periwinkle are not as toxic as the Oleander, or some of the Dogbanes that grow further north. It is well to be careful of all plants that have a milky juice as many of them contain a poisonous agent. For example, many of the milkweeds are dangerous to cattle if they graze upon them.

Castor Oil (Ricinus communis). Herb up to 20 ft. tall; much branched; thick, smooth, fleshy, dark red stems; large leaves deeply cut into 7-9 lobes with toothed edges, green on top, purplish red beneath; bears incomspicuous flowers followed by clusters of small balls covered with soft, fleshy, red spines, which split open to reveal 3 glossy, black or mottled seeds. All parts of plant contain ricin, a blood poison, and children should be warned not to eat the seeds. The oil from these seeds, with the ricin removed is Castor Oil.

Foxglove or Fairy Fingers (Digitalis purpurea) Spikes of many tubular, bell-like flowers, 1″ long, white, rose, pink or lavender; leaves very slender with sharply toothed edges. Long a garden favorite, this beautiful pant is dangerous. The leaves and roots yield the powerful drug, digitalis, widely used in heart medicines.

Larkspur (Delphinium caroliniamum). Palmate grayish-green leaves, divided into 5 deeply and finely cut leaflets; racemes of blue, purplish or white flowers are borne at top of slender erect stalk. This loved and familiar garden flower contains several toxic alkaloids, the most powerful being delphinine, which are extracted from the seeds and used in medicines. In the west, Larkspur poisoning of cattle is a serious problem.

20

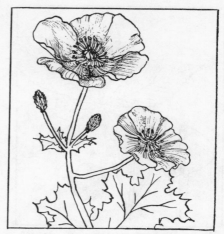

Pokeweed (Phytolacca americana). Herbaceous; 4-10 ft. tall; stout, fleshy stems, often purple tinged; leaves flat, broader below the middle, 3-10 in.; erect racemes of small, white flowers, followed by dark reddish-purple, juicy berries. The young shoots are cooked in two waters and eaten as greens, but the thick, fleshy roots, and the seeds of the berries, contain a dangerous poison, which causes paralysis of the respiratory organs. The root has been used extensively in the preparation of certain drugs.

Prickly or Mexican Poppy (Argemone mexicana). 1-3 ft. tall; stout stems; leaves 3-6 in. alternate, coarsely lobed, spiney; large single flowers, yellow, 2-4 in. across. Fruit a prickly capsule, with many seeds. In the unripe capsule is a milky juice containing opium, from which is derived laudanine, morphine, etc. Argemone Alba has white flowers.

Rhubarb or Pie Plant (Rheum Rhaponticum). This common, perennial, with very large leaves and fleshy, reddish stalks, is grown in gardens. The acid leaf stalks are cooked and eaten in pies or as sauce, but the leaf blades are poisonous. Several deaths have been reported from people eating the cooked leaf blades.

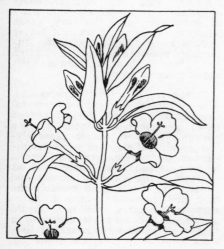

Yellow Jessamine (Gelsemium sempervirens). This beauty is a climbing vine; opposite, ovate, pointed, glossy, green leaves; flowers yellow or white, 1-1½ in. long; in small clusters, along the slender, woody stem, fragrant. This vine climbs in profusion over shrubs and trees along highways, and in woods and thickets. All parts of the plant are toxic, and the roots yield a powerful drug. It is claimed that even birds and bees avoid it.

Red Buckeye (Aesculus pavia). Small tree or shrub; leaves opposite, palmate, with 5-7 serrate leaflets; red flowers in erect panicles; fruit a leathery capsule, containing 1-3 large, shiny, brownish seeds. Young shoots and leaves especially poisonous and children have been poisoned from eating the seeds.

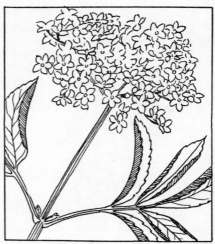

Elder (Sambucus simpsonii). Shrub or small tree, to 15 ft. tall; pinnate leaves; leaflets 3 in. long, pointed, toothed; flat heads 4-8 in. across, of tiny, white flowers; followed by clusters of small black berries, which can be cooked, and eaten in pies or as jam, but be careful, for all the rest of the plant is poisonous. Children have been made ill from chewing on the bark, and cattle have eaten leaves and young shoots with fatal results.

Boxwood (Buxus sempervirens). Shrub used for hedges; many branches; covered densely with pairs of round-oval, dark, leathery, green leaves ¼ in. long. Bark and leaves contain several toxic alkaloids. Bitter taste prevents animals or children from chewing them, so Boxwood isn't apt to do much harm.

Chinaberry Tree (Melia azedarach). A well known door-yard ornamental tree; 20-40 ft. high; thick, furrowed bark; an umbrella-shaped, d e n s e crown of leaves; leaves pinnately-compound; leaflets dark green, pointed, toothed; purple, fragrant flowers, in long-stalked, loose panicles; fruit a drupe, ½ in. diameter, smooth and yellow. The fruits (and possibly the bark and flowers) contain a toxic, narcotic substance. Children should be warned not to put the drupes in their mouths, and chickens and livestock kept away from the trees.

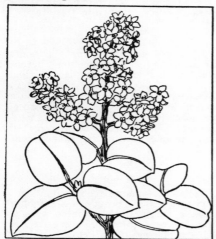

Wax Privet (Ligustrum lucidum). Foliage thick; leaves dark green, glossy, evergreen, ovate, pointed; tiny, white, fragrant flowers, borne in dense, terminal clusters; black berries; widely used for hedges. Berries and leaves poisonous.

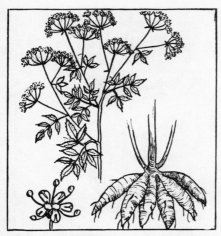
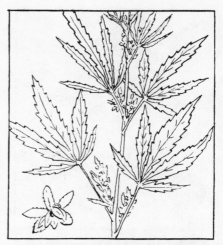

Water Hemlock, Wild Parsnip or Spotted Cowbane (Cicuta maculata). Herbaceous, 2½-7 ft. tall; erect, slender, smooth stalks are purple spotted; leaves alternate with 9 or more slender, pointed leaflets with toothed edges; small, whitish flowers in loose, terminal clusters; grows in damp, marshy places. This is one of the most poisonous species in the U. S. All parts of the plant are poisonous, but the roots are the most deadly. People sometimes mistake them for wild parsnips. Be careful. If eaten, a small amount is fatal.

Precatory Bean, Rosary Pea (Abrus precatorius). Pea or Bean Family; naturalized vine; stem woody; leaves compound, pinnate, 24 ovate leaflets; dense, racemes of pink or lavender flowers; ½ in. long; broad pods 1½ in. long; small red and black seeds. These seeds are extremely poisonous. Fortunately they are hard coated, and if swallowed without being chewed, not much of the poison will be absorbed.

Macaw Tree (Daubentonia punicea). ▶ One of the Pea Family. Small tree or shrub; 10-12 ft high; low-spreading, with many branches; orange-red, pea-shaped flowers in drooping clusters; seed pods 3-4 in. long, with 4 crinkly wings running the length of the pod, which dries to a dark brown. Seeds are very hard and contain the poison saponin.

Marijuana, Indian Hemp or Hashish (Cannabis sativa). Coarse, rough-stemmed annual, 3-9 ft. tall; palmate leaves, with 5-7 narrow, toothed leaflets; flowers small, green, in axillary, spike-like clusters. The plant contains at least three powerful narcotics, which are more potent when it is grown in a warm climate. Grows locally as a weed of waste places from coast to coast, and sometimes as a crop. The upper leaves and flower bracts are dried and smoked by drug adicts, in pipes or cigarettes.

23

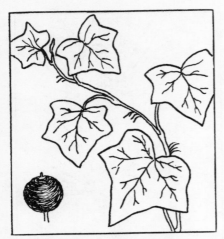

English Ivy (Hedera Helix). A vine, well known as an ornamental house plant, and often planted where it can climb as a cool, green cover over walls and buildings. Stem woody; leaves alternate, evergreen, shallow pointed lobes, with veins lighter than the leaf; small greenish flowers; fruit a small black berry. The entire plant contains the glucose hederin, but the leaves and berries are especially toxic. Some people develop a severe dermatitis from handling the ivy.

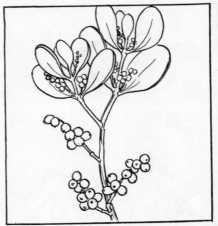

Mistletoe (Phoradendron flavescens). Round-oval, thick, fleshy, smooth, green leaves, 1 in. long; small, waxy, whitish berries. A parasite which takes for its own growth, the life blood of the tree upon which it lives. Great clumps of Mistletoe can be seen, high in the branches of trees. The berries are dangerously poisonous, if eaten.

Atamasco Lily   Star-of-Bethlehem

Amaryllis (Amaryllis). Several varieties of Amaryllis are seen in Florida — gorgeous, huge, red trumpet-shaped flowers — or shading to white. Bulbs are poisonous, and should not be left where children can get at them.

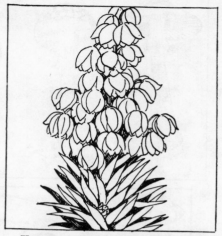

Sandbur or **Sandspur** (Cenchrus pauciflorus). Grass-like plant with long, reclining stems, bearing at the ends clusters of small burs, whose tiny spines have recurved tips, and may break off in the skin. Remove with tweezers if possible, but do not dig for one that is deeply embedded. It will fester and come out easily, and disinfectant can be applied.

Star-of-Bethlehem (Ornithogalum umbellatum). Slender, ornamental plant, from Europe, escaped here as a weed in southeastern states. From a bulb, spring slender, grass-like leaves, and a single erect stem, bearing at the top a branched cluster of dainty, six-petalled, star-like flowers, white with a green vein on the back. All parts of the plant are poisonous whether fresh or dried. The flowers keep for weeks without water, and are sold as ornamentals, with the bulbs. They should be kept out of children's reach.

Atamasco-lily, Rain-lily, Zephyr-lily (Zephyranthes atamasco). A few narrow, grass-like leaves at the base of single, erect stalk. At the top of stalk is a lovely, solitary, lily-like flower, 2-3 in. long, pale pink or white, becoming tinged with purple as the flower ages. They spring up after rains, in ditches, and any low ground. They grow from a small bulb which is poisonous.

Yucca or Spanish Bayonet (Yucca alfoifolia). Thick spike, to 8 ft. tall of stiff, sword-like, very sharp-pointed leaves; large spike of waxy-white, drooping, bell-shaped flowers. Be careful of our beautiful Yuccas, for the needle-like points can inflict painful wounds.

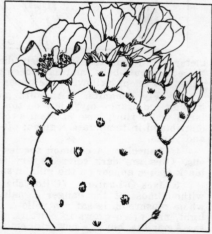

Prickly Pear (Opuntia). Semi-erect, in clumps; 2-5 ft. high; with many branches of flat, pad-like, spiny joints; lovely, large, lemon yellow flowers. These cactus are armed with spines from small ones to some 1 in. long.

25

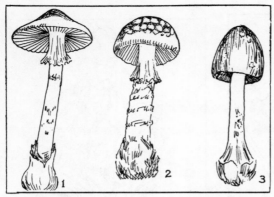

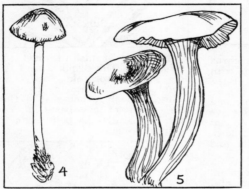

Mushrooms and Toadstools are responsible for more deaths than any other plant. Many kinds are edible, and d e l i c i o u s, others are as d e a d l y as a rattlesnake. Danger signals are: a cup at the base of the stem; a ring around the stem; white gills; any mushroom that turns bluish when broken; any mushroom that exudes a milky juice when broken, unless the juice is reddish. Some edible m u s h r o o m s have one or more of these danger signs, so it is very difficult to tell them apart. There are several w i d e - spread, but *False* beliefs concerning mushrooms, such as: only edible ones will peel easily; only brightly colored ones are poisonous; only e d i b l e s p e c i e s grow in c l u m p s ; poisonous fungi taste bitter; cooking will destroy the poison. So, BE CAREFUL! It is *dangerous* to gather edible fungi unless you are an expert.

1 and 3 are the **Death Cup** or **Death's Angel** (Amanita phalloides). Glistening white gills, a ring and a death cup; peels easily; has a pleasant flavor; cap is 3-4 in.; diameter of various coloring widely distributed. Very young Amanitas look like the edible puff-balls, and have been mistaken for them with fatal results. One variety is lemon-yellow on the upper cap surface. No. 3 is the young Amanita.

2. **Fly Amanita.** (Amanita muscarius). Large mushroom; caps 3-8 in. in dia.; red-orange or yellow on top with white or yellowish warts; gills white; large white ring; base of stem swollen and rough, but no distinct cup; widely distributed in temperate regions; often found growing near pine trees, pastures, and yards.

4. **Paneolus.** A common species that is innocent in appearance, but poisonous. Gills are dark colored; very slender stems, greyish or brownish; nearly black spores appear on the gills in spots.

5. **Jack-O-Lantern.** (Clitocybe illudens). A large, conspicuous variety, without most of the danger signals. The caps are to 8 in. across, in color the whole mushroom is saffron yellow. It glows in the dark with a phosphorescent light, and often grows in clusters. Pleasant to taste, it is not as poisonous as the Amanitas, but is a powerful emetic.

Green-gilled Lepiota (Lepiota morgani). Large, poisonous mushroom common in Florida; green gills, with course scales on upper surface of cap; ring near top of stem; pastures, open woods, lawns.

Symptoms from eating poisonous mushrooms are severe pains in the abdomen, 6-15 hours after eating, weakness, coma and death in 3-8 days, if sufficient poison was taken. Always pull up all toadstools and mushrooms that spring up in your yard. I have given only a few of the poisonous varieties.

Other Poisonous Plants

A list of poisonous ornamental plants, found in the southeastern states, but not considered as Florida flowers:

Lily-of-the-Valley	Rhododendron
Bloodroot	Monkshood or Aconite
Moonseed	Lady Slipper (dermatitis)
Laurel	Belladona
Matrimony Vine	Dutchman's Breeches
Christmas Rose	Corn-cockle
May-apple	Bittersweet
Daphne	Meadow Saffron
Buttercup	Red, and White Banberry
Foxglove	Holly Tree (berries)

Acknowledgments

I wish to thank Mr. Erdman West, Professor of Botany, with the College of Agriculture, University of Florida, and Botanist and Mycoligist with the Florida Agricultural and Experiment Station, for checking the section on Poisonous Plants. I also wish to thank the editor of Florida Wildlife Magazine, Mr. Bill Hansen, for having his staff check the section on Poisonous snakes.

Various books have been of invaluable help to me in preparing this small book, among them Muenscher's *Poisonous Plants in the United States*; Greene's and Blomquist's *Flowers of the South*; the bulletin *Poisonous Plants in Florida* by Erdman West and M. W. Emmel; Ditmar's *Nature Library, Book of Reptiles; The Animal Kingdom Book of Reptiles, Book of Insects* and articles in Florida Wildlife, National Geographic and Nature Magazines.

Poisonous Plants

Page

Page

Index
Snakes - Insects - Marine Life

Poisonous Plants